From Earth to the Stars:

A Guide to the Milky Way for the Curious

SUMMARY

CHAPTER 1: INTRODUCTION TO THE MILKY WAY 5

brief history of the discovery of the milky way 5

CHAPTER 2: STRUCTURE AND COMPOSITION OF THE MILKY WAY 10

Description of the General Shape of the Milky Way 10

CHAPTER 3: STARS AND STAR SYSTEMS IN THE MILKY WAY 15

Types of Stars Found in the Milky Way 15

CHAPTER 4: NEBULAE AND STAR FORMATION REGIONS IN THE MILKY WAY 21

Exploration of Nebulae in the Milky Way 21

CHAPTER 5: BLACK HOLES, NEUTRON STARS AND QUASARS 27

Exotic Wonders in the Milky Way 27

CHAPTER 6: PLANETARY SYSTEMS AND EXOPLANETS IN THE MILKY WAY: EXPLORING NEW WORLDS 32

Discoveries of Planets Outside Our Solar System 32

CHAPTER 7: HISTORY AND EVOLUTION OF THE MILKY WAY: A COSMIC JOURNEY THROUGH TIME 39

The Formation of the Milky Way 39

CHAPTER 8: SPACE MISSIONS AND OBSERVATORIES: UNCOVERING THE MYSTERIES OF THE MILKY WAY 45

Space Missions that Study the Milky Way 45

CHAPTER 9: LIFE IN THE MILKY WAY: EXPLORING POSSIBILITIES BEYOND EARTH 52

The Search for Extraterrestrial Life 52

CHAPTER 10: CURIOSITIES AND MYTHS: EXPLORING THE CULTURAL NARRATIVES AND INTRIGUING MYSTERIES OF THE MILKY WAY... 59

Mythology and Cultural Narratives:...59

CHAPTER 11: PHOTOGRAPHS AND SPACE IMAGES: A VISUAL JOURNEY THROUGH THE MILKY WAY 66

Exploring the Milky Way through the Cosmic Lens................................66

CHAPTER 12: CHALLENGES AND UNSOLVED MYSTERIES:..78

The Unexplored Frontiers of Cosmic Understanding..........................78

CHAPTER 13: SCIENTIFIC AND CULTURAL IMPORTANCE OF THE MILKY WAY: A JOURNEY OF DISCOVERY AND INSPIRATION 87

Scientific Contributions:...87

CHAPTER 14: CONCLUSION - NAVIGATING THE MYSTERIES OF THE MILKY WAY 93

CHAPTER 1: INTRODUCTION TO THE MILKY WAY

BRIEF HISTORY OF THE DISCOVERY OF THE MILKY WAY

The Milky Way, a majestic band of light that paints the night sky, has fascinated humanity for centuries. Its discovery and understanding involve an intriguing journey, permeated by myths, empirical observations and scientific advances. Throughout history, humanity has looked to the firmament with questions, triggering a search to understand our position in the cosmos.

The beginnings of celestial exploration date back to antiquity, when civilizations such as Babylon and ancient Greece began recording constellations and observing patterns in the night sky. The Milky Way, often called the "Milk Path" by various cultures, was seen as a vast luminous highway that connected heaven to earth. Myths and legends were woven around this celestial strip, each carrying a unique interpretation.

However, it was only in the era of systematic astronomical observation that the mysteries of the Milky Way began to be unraveled. With the invention of the telescope by Galileo Galilei in the 17th century, the doors of cosmic exploration were opened. Galileo, when pointing his telescope at the night sky, realized that the Milky Way was made up of a multitude of individual stars, previously invisible to the naked eye. This revelation transformed the vision of a luminous road into an immense collection of distant celestial bodies.

In the 18th century, British astronomer William Herschel played a crucial role in understanding the structure of the Milky Way. Using increasingly powerful telescopes, Herschel mapped the distribution of stars and proposed a disk-shaped model for the galaxy. His pioneering vision of galactic structure laid the foundation for modern understanding of the Milky Way.

Fast forward to the 20th century, with the advent of astrophotography and space observatories such as the Hubble Space Telescope, astronomers were able to delve even deeper into the galaxy's rich details. Surprising discoveries, such as the presence of supermassive black holes at the center of the Milky Way, have expanded our understanding of the dynamics and complexity of the galaxy we call our cosmic home.

Contextualization Within the Known Universe

The Milky Way is not an isolated entity in the vast cosmic ocean; she is part of an intricately interconnected universe. In terms of scale, our galaxy is one of billions, a small piece of a still largely unsolved cosmic puzzle. Neighboring galaxies, like Andromeda, dance in interstellar space, each with its own unique history and composition.

Furthermore, the Milky Way is not just a static collection of stars, but a dynamic system in constant evolution. Gravitational interaction with neighboring galaxies, star formation in their spiral arms, and extraordinary cosmic events continually shape the fate of our galaxy. It is in this cosmic context that we must understand the grandeur and beauty of our galactic home.

As we delve deeper into the exploration of this ebook, the extraordinary cosmic spectacle that is the Milky Way will be revealed. From the star systems at its core to the vast star-forming regions in its spiral arms, each chapter will provide a clearer view of the complexity and beauty that permeates our galaxy.

On this journey, we will embark on a journey through time and space, unlocking the secrets of the Milky Way and exploring the wonders it hides deep within the cosmos. Prepare yourself for a cosmic odyssey that will expand the limits of your understanding of the vastness of the universe.

CHAPTER 2: STRUCTURE AND COMPOSITION OF THE MILKY WAY

DESCRIPTION OF THE GENERAL SHAPE OF THE MILKY WAY

The Milky Way, that vast band of light that adorns the night sky, is much more than a simple cosmic decoration. Its structure, complex and majestic, is the result of billions of years of cosmic evolution. When we observe the Milky Way from our terrestrial perspective, we see a band of diffuse light, but its true shape is revealed when we delve deeper into astronomical exploration.

The Milky Way is, in essence, a spiral galaxy. Imagine a flat disk slowly rotating in space, with spiral arms extending from a central core. This is the fundamental architecture of our galaxy, a cosmic masterpiece that defies the imagination. It is crucial to understand this form to appreciate the complexity of its individual components.

Main Components: Bulb, Disc, Halo and Spiral Arms

1. Bulb:

At the heart of the Milky Way, we find a central component known as the galactic bulge. This is a dense cluster of stars, gas and dust concentrated around the galactic core. The bulge is a dynamic, activity-filled region where stars orbit around a supermassive black hole. The density of stars in this domain creates a distinct luminosity when observed from afar, and understanding it is critical to unlocking the mysteries at the heart of the galaxy.

2. Disc:

The disk of the Milky Way is a vast, flat region that extends beyond the bulge. This is the component most visible from our position on the edge of the galaxy. In it, we find most of the stars, planets, interstellar gas and cosmic dust that make up our galaxy. The stars in the disk orbit the galactic center, moving in elliptical paths over millions and billions of years.

3. Halo:

In addition to the disk, the Milky Way is surrounded by a vast region called the galactic halo. This is a less dense domain, composed mainly of old stars, diffuse gas and dark matter. The halo extends considerable distances beyond the disk and plays an important role in the long-term stability and evolution of the galaxy.

4. Spiral Arms:

One of the most distinctive aspects of the Milky Way is its spiral arms, curvilinear patterns of stars, gas and dust that extend outward from the disk. These arms are places of intense star formation activity, where new stars are born from clouds of gas and cosmic dust. Spiral arms contribute to the visual beauty of the galaxy and play a vital role in its dynamics.

When we look at the Milky Way, we are not just witnessing a distant structure; we are observing cosmic evolution in action. Each component, from the bulb to the spiral arms, contributes to the cosmic harmony that defines our galaxy. It is a masterpiece of cosmic proportions, a stellar tapestry that connects us to the vast universe beyond our night vision. Knowing the structure and composition of the Milky Way is essential to understanding our position in the cosmos and appreciating the complexity and beauty it offers to those who venture to explore it.

CHAPTER 3: STARS AND STAR SYSTEMS IN THE MILKY WAY

TYPES OF STARS FOUND IN THE MILKY WAY

The Milky Way, our galactic home, is a vast stellar kingdom that is home to an incredible diversity of stars, each with its own unique characteristics. The classification of stars is based on several factors, such as temperature, luminosity, size and chemical composition. As we explore the galaxy, we find stars ranging from red giants to white dwarfs. This stellar variety creates a fascinating celestial mosaic.

1. Main Sequence Stars:

The vast majority of stars in the Milky Way, including our Sun, belong to the category of main sequence stars. These stars are in a phase where they convert hydrogen into helium through nuclear fusion. The Sun, for example, is a main sequence star of spectral type G, a yellow star of moderate temperature.

2. Red Giants and Supergiants:

As stars age, many of them expand to become red giants or, in more extreme cases, supergiants. These stars are impressive in size and often exhibit variations in their luminosity. Betelgeuse, located in the constellation Orion, is a notable example of a red supergiant.

3. White Dwarfs:

When stars with a mass similar to or less than the Sun run out of nuclear fuel, they can transform into white dwarfs. These are extremely dense stars, remnants of the collapsed cores of ancient stars. The white dwarf Sirius B, companion to the star Sirius A, is a classic example of this type of star.

4. Binary Stars:

In the Milky Way, many stars are in binary systems, where two stars orbit around a common center of mass. These systems can be of two main types: visual binaries, where stars can be distinctly observed, and spectroscopic binaries, where stars are identified through changes in their spectra.

5. Multiple Star Systems:

In addition to binaries, there are stellar systems that involve three or more stars. These multiple systems can have a variety of configurations, including gravitationally bound stars in triple, quadruple, or even more complex systems. Alpha Centauri, a triple system, is a nearby example in the solar neighborhood.

Star Systems, such as Binaries and Multiple Systems

1. Binary Stars:

Binary stars are fascinating systems in the Milky Way, representing a considerable portion of the stellar population. The proximity between stars in a binary system can vary, from stars that share a close orbit to those that are separated by vast distances. Visual binaries are easily discernible, while spectroscopic binaries require spectral analysis to identify the presence of invisible companions.

2. Multiple Systems:

The complexity of multiple star systems in the Milky Way is astonishing. Stars orbiting a binary star, for example, form triplet systems, adding an additional layer of gravitational dynamics. In addition, there are stars that group together in even more elaborate configurations, involving four, five or more stellar members interacting gravitationally.

3. Stellar Evolution in Binary Systems:

The presence of a companion star can significantly influence stellar evolution. In close binary systems, one star can transfer mass to another, resulting in events such as new supernova explosions. Gravitational interaction in multiple systems can lead to unstable orbits or even the ejection of stars from the system.

4. Importance in Planetary Formation:

Star systems with planets are also common in the Milky Way. In some cases, the presence of companion stars can influence the formation and orbit of planets around a primary star. The complex dynamics of these star systems create an intriguing environment for studying planetary formation and habitability.

Exploring the variety of stars and star systems in the Milky Way is an exciting journey through the cosmic realm. Every star, every binary system and every multiple system tells a unique story of formation and evolution. By delving deeper into understanding these elements, we gain a more complete and fascinating view of the dynamics that permeate our galaxy. Exploration continues to reveal stellar secrets that expand our understanding of the universe in which we live.

CHAPTER 4: NEBULAE AND STAR FORMATION REGIONS IN THE MILKY WAY

EXPLORATION OF NEBULAE IN THE MILKY WAY

Nebulae, cosmic clouds of interstellar gas and dust, play a crucial role in stellar evolution and in creating environments suitable for the birth of new stars. In the vastness of the Milky Way, a variety of nebulae present distinct shapes and characteristics, each contributing to the beauty and complexity of the cosmic landscape.

1. Emission Nebulae:

Emission nebulae are regions where gas atoms, often hydrogen, are excited by radiation from neighboring stars and emit visible light. These nebulae often glow with vibrant colors, such as the characteristic red of the Lagoon Nebula (M8) and the pink of the Crab Nebula (M1).

2. Dark Nebulae:

In contrast to emission nebulae, dark nebulae are dense regions of interstellar dust that block light from background stars. These nebulae are often observed as dark areas against the stellar background, creating fascinating silhouettes. The Horsehead Nebula is a notable example of a dark nebula.

3. Planetary Nebulae:

Planetary nebulae are formed when a Sun-like star reaches the final stage of its life and expels outer layers of expanding gas. The remaining nebula can acquire spherical or complex shapes. The Ring Nebula (M57) is a classic example of a planetary nebula.

4. Reflection Nebulae:

Reflection nebulae are formed when light from nearby stars is reflected by dust grains. These nebulae often appear blue due to the preferential scattering of blue light. The Monkey Head Nebula is an intriguing example of a reflection nebula.

5. Gas and Dust Nebulae:

Diffuse nebulae of gas and dust are sites of intense star formation activity. The Orion Nebula, one of the most famous in the Milky Way, is a stunning example. These nebulae are often associated with young star clusters, where new stars are born from the surrounding nebulous material.

Star Formation Regions and the Orion Nebula

1. Orion Nebula (M42):

The Orion Nebula is one of the most spectacular star-forming regions in our galaxy. Located in the Orion constellation, it is visible to the naked eye and presents incredible visual complexity when observed with telescopes. This nebula is an active region of star formation, home to massive young stars, stellar clusters, and pillars of gas and dust. The presence of the famous Trapezium, a group of four young and luminous stars, highlights the vitality of this region.

2. Young Star Clusters:

Within the Orion Nebula, young star clusters such as the Trapezium are evidence of the dynamic evolution of this region. These clusters are formed from the same nebulous material, and the newborn stars contribute to the nebula's characteristic luminosity.

3. Gas and Dust Pillars:

The Orion Nebula is also known for its impressive pillars of gas and dust, highlighted in iconic images captured by space telescopes like Hubble. These pillars are environments suitable for star formation, where the gravitational compression of material leads to the creation of new stars.

4. Stellar Life Cycle:

By studying regions like the Orion Nebula, we gain valuable insights into the stellar life cycle. From the formation of young stars to the eventual explosion of supernovae and the dispersal of elements into space, this region serves as a cosmic laboratory, allowing us to trace the different phases of stellar evolution.

Scientific and Aesthetic Importance

In addition to their scientific importance, nebulae and star-forming regions capture the imagination and inspire aesthetics. They are a visual testament to the cosmic dynamics that give rise to celestial wonders. The Orion Nebula and its counterparts throughout the Milky Way not only illuminate the night sky, but also illuminate our understanding of the ever-evolving universe. Each nebula tells a unique cosmic story, shaped by the interaction between stars, gas and dust over billions of years. This cosmic narrative continues to unfold, as we explore the skies and depths of the Milky Way in search of yet-to-be-revealed secrets.

CHAPTER 5: BLACK HOLES, NEUTRON STARS AND QUASARS

In the cosmic vastness of the Milky Way, beyond the stars and nebulae, we find exotic objects that defy conventional understanding. Black holes, neutron stars and quasars are extraordinary cosmic entities, each containing fascinating mysteries and contributing to the richness of the stellar landscape.

1. Black Holes: Portals to the Unknown

Black holes, regions of space where gravity is so intense that nothing, not even light, can escape, are some of the most enigmatic entities in modern cosmology. Formed by the gravitational collapse of massive stars, these objects are invisible to traditional observation methods, being detected mainly through their gravitational effects on neighboring objects.

Black holes can be classified into stellar black holes, formed by the death of massive stars, and supermassive black holes, found at the centers of galaxies, including the Milky Way. The supermassive black hole known as Sagittarius A* resides at the center of our galaxy, exerting a powerful gravitational pull.

Although black holes absorb light and become virtually invisible, they can be detected indirectly by the radiation emitted by the surrounding gases as they are consumed by the black hole. Exploring these dark cosmic regions opens a window into the unknown and sheds light on the extremes of theoretical physics.

2. Neutron Stars: Density Beyond Imagination

Neutron stars are the collapsed remnants of massive stars that have gone supernova. These incredibly dense objects are composed primarily of neutrons, resulting from the extremely high pressure caused by gravitational collapse. The density in a neutron star is so great that a teaspoon of its matter would weigh billions of tons on Earth.

The rotation of neutron stars can accelerate dramatically after they collapse, giving rise to pulsars, neutron stars that emit pulsating radiation toward Earth. These exotic objects are excellent markers of cosmic time and provide a unique insight into the extremes of nuclear physics.

3. Quasars: Distant Lighthouses of the Cosmos

Quasars, or "quasi-stellar sources," are active, extremely luminous galactic nuclei powered by supermassive black holes.

These objects emit radiation in a wide range of wavelengths, from radio to gamma rays, making them detectable at cosmic distances.

Quasars play a crucial role in understanding the evolution of galaxies and the early universe. They are observed at such great distances that they provide a glimpse into very ancient epochs of the cosmos, when galaxies were in the early stages of formation. Studying these cosmic beacons not only reveals the mysteries of supermassive black holes, but also sheds light on the evolutionary dynamics of the universe over time.

The Cosmic Dance of Exotic Stars in the Milky Way

The presence of black holes, neutron stars and quasars in the Milky Way reveals a complex and dynamic cosmic dance. These objects, although exotic, are intrinsic parts of the cosmic fabric that shapes the evolution of our galaxy. Studying these entities not only expands our understanding of fundamental physics, but also allows us to delve into the extreme and fascinating conditions present in many corners of the Milky Way.

Continued exploration of these cosmic wonders not only challenges our current knowledge, but also inspires future discoveries and deepens our understanding of the universe. Every black hole, neutron star and quasar in the Milky Way is a unique piece in the great cosmic puzzle, inviting us to contemplate the limits of physical reality and the intrinsic beauty of the exotic nature that permeates interstellar space.

CHAPTER 6: PLANETARY SYSTEMS AND EXOPLANETS IN THE MILKY WAY: EXPLORING NEW WORLDS

DISCOVERIES OF PLANETS OUTSIDE OUR SOLAR SYSTEM

The fascination with the search for life beyond Earth has led astronomers to turn to the vast cosmos in search of planets outside our solar system, known as exoplanets. The journey of discovery began to gain prominence in recent decades, resulting in a revolution in understanding the diversity and abundance of planetary systems in the Milky Way.

1. The Exoplanet Revolution:

Until the mid-20th century, the idea of finding planets beyond our solar system was more science fiction than reality.

However, in recent decades, technological advances and innovative observation techniques have opened up new perspectives. The discovery of the first exoplanet orbiting a Sun-like star, 51 Pegasi b, in 1995 marked the beginning of a revolution in astronomy and triggered an intensive search for new worlds.

2. Exoplanet Detection Methods:

Several methods have been developed to detect exoplanets, each with its own advantages and challenges. The transit method observes the temporary decrease in a star's luminosity when a planet passes in front of it. Another method, radial velocity, measures variations in the star's speed caused by the gravitational pull of a planet. Newer methods, such as direct imaging, capture direct images of exoplanets.

3. Diversity of Exoplanets:

Exoplanet discoveries have revealed surprising diversity in terms of size, composition and orbit. We find everything from gas giants similar to Jupiter to small rocky planets, some orbiting multiple stars or binary systems. The variety of planetary conditions challenges previous expectations and stimulates the search for potentially habitable planets.

Notable Planetary Systems in the Milky Way

1. TRAPPIST-1:

One of the most intriguing planetary systems discovered so far is TRAPPIST-1, an ultracool dwarf star with seven known exoplanets. The system gained prominence due to the possible habitability of some of its planets. Studies have suggested that some of these worlds could have liquid water, a key ingredient for life as we know it.

2. Kepler-90:

The Kepler-90 system is a notable example of a planetary system that rivals our own. With eight confirmed planets, Kepler-90 challenges previous assumptions about the distribution of planets around distant stars. Confirmation of these distant worlds opens a window into understanding the complexities of planetary formation.

3. WASP-121:

WASP-121 is a planetary system that highlights the extreme diversity of conditions found outside our solar system. One of the planets in this system, WASP-121b, is a gas giant with extreme temperatures, reaching levels never before observed. Studying these extreme environments expands our understanding of variability in planetary systems.

4. HD 219134:

The HD 219134 system is home to at least four exoplanets. One of the planets, HD 219134b, is a rocky planet that orbits very close to its parent star. These findings highlight the diversity of planetary architectures and challenge traditional predictions about the distribution of planet types in stellar systems.

5. Proxima Centauri b:

Proxima Centauri b orbits the closest star to our solar system, Proxima Centauri, in the constellation Centaurus. This exoplanet is a rock with a mass similar to Earth, and its proximity to the star suggests the possibility of suitable conditions for the existence of liquid water, increasing interest in the search for signs of life.

Scientific and Philosophical Importance

The search for exoplanets is not just a scientific exploration, but also a philosophical journey that makes us question our place in the universe. The discovery of planets in habitable zones, where liquid water may exist, fuels the dream of finding signs of life beyond Earth.

Furthermore, the diversity of observed planetary systems suggests that our own solar system may be just one of many possible configurations. Continued exploration of these distant systems expands our horizons and invites us to contemplate the variety of possible worlds.

As we investigate planetary systems and exoplanets in the Milky Way, we delve into a cosmic ocean of possibilities. Each new world discovered not only increases our understanding of planetary formation, but also takes us one step closer to answering the fundamental question: are we alone in the universe? Exploring these new worlds, even at unimaginable distances, connects us to the vast cosmic panorama that is the Milky Way, providing a unique and inspiring view of our position in the cosmos.

CHAPTER 7: HISTORY AND EVOLUTION OF THE MILKY WAY: A COSMIC JOURNEY THROUGH TIME

THE FORMATION OF THE MILKY WAY

The story of the Milky Way is a cosmic narrative that unfolds over billions of years. Its origin dates back to the beginnings of the universe, when the cosmos was a vast field of cosmic gas and dust, permeated by density fluctuations that, over time, led to the formation of the first galaxies.

The Milky Way is a spiral galaxy, meaning it has a flat, rotating disk with distinct spiral arms. Its formation began with the agglomeration of gas and dark matter in denser regions of the early universe. Gravity acted as the cosmic architect, assembling this material into increasingly massive structures, eventually giving rise to the protodisk that would become the Milky Way.

Training Phases:

During the early stages, the formation of the Milky Way involved the merger of small galactic structures, combining gas, stars and dark matter in a gradual process. As the galactic disk developed, star formation intensified in the spiral arms, creating the foundations for the stars that now dot the night sky.

Stellar Evolution and Chemical Elements

1. Chemical Enrichment:

The process of star formation in the Milky Way has played a vital role in the chemical enrichment of the interstellar medium. Stars, as they go through their life cycle, synthesize heavier elements through nuclear reactions. When these stars eventually explode as supernovae, they release these elements into space, enriching interstellar gas with a variety of elements.

2. Generation of Elements Necessary for Life:

The Milky Way's progressive chemical enrichment is crucial to the formation of planetary systems and the emergence of life. Elements such as carbon, oxygen, nitrogen and iron, essential for life as we know it, were forged in stellar furnaces and dispersed by supernovae throughout the cosmic eons.

Important Events in the History of the Milky Way.

1. Galactic Mergers:

Throughout its history, the Milky Way has participated in gravitational interactions with other galaxies. Galactic mergers are significant events that shape the structure and dynamics of the galaxy. Stars, gas and dark matter from smaller galaxies were incorporated into the Milky Way, contributing to the complexity of its formation.

2. Formation of Spiral Arms:

The spiral arms, striking features of the Milky Way, were formed due to the galaxy's rotational dynamics and gravitational interactions with other galaxies and clouds of interstellar matter. These arms are places of intense stellar activity, with the formation of new stars from clouds of gas and dust.

3. Galactic Halo and Stellar Population:

Beyond the galactic disk, the Milky Way is surrounded by a halo of ancient stars. These stars, which date back to the galaxy's earliest days, provide clues about the events that shaped its evolution. The halo's stellar population is a testament to the Milky Way's cosmic past.

4. Activity in the Galactic Core:

The core of the Milky Way is home to a supermassive black hole called Sagittarius A*. Periodic accretion events of gas and dust by the central black hole result in intense radiation emissions. This activity in the galactic core is an intrinsic part of the Milky Way's dynamic history.

5. Formation of Star Clusters:

Throughout its history, the Milky Way has generated star clusters, dense concentrations of stars born from the same material. These clusters are archaeological records of star formation at different times in galactic history.

The Future of the Milky Way

The study of the history and evolution of the Milky Way is an ever-evolving pursuit. As observational technologies improve and new discoveries are made, we continue to advance our understanding of how our galaxy formed and evolved over cosmic time.

Looking to the future, the Milky Way still has many stories to tell. Investigating cosmic events, galactic mergers, ongoing star formation, and the dynamics of the galactic core will open new chapters in our understanding of the ongoing evolution of this cosmic spiral we call home. Every star, every supernova, every cosmic event is a page in the history of the Milky Way, a story we continue to unravel as we explore the mysteries of our ever-changing galaxy.

CHAPTER 8: SPACE MISSIONS AND OBSERVATORIES: UNCOVERING THE MYSTERIES OF THE MILKY WAY

Space missions have played a crucial role in exploring and understanding the Milky Way. Over the decades, space probes and orbiting observatories have contributed significantly to our view of the cosmos and deepened our knowledge of the galaxy in which we live.

1. Gaia: Mapping the Milky Way with Unprecedented Accuracy

The Gaia mission, launched by the European Space Agency (ESA), has been one of the most revolutionary missions to study the Milky Way. Since its launch in 2013, Gaia has aimed to map the position and movement of more than a billion stars in our galaxy. This space probe is providing an unprecedented three-dimensional view of the Milky Way's stellar structure, providing fundamental data for understanding stellar distribution, galactic dynamics and historical star formation.

2. Hubble: Revealing the Wonders of the Milky Way.

The Hubble Space Telescope, operated by NASA and ESA, has been an icon of astronomy since its launch in 1990. While its orbit around the Earth allows for clear observations without atmospheric interference, Hubble has contributed significantly to our understanding of the Milky Way . Its spectacular images reveal incredible details of nebulae, star clusters, black holes and other celestial objects within our galaxy.

3. Kepler: Discovering Exoplanets in the Milky Way

Although the Kepler mission did not directly target the Milky Way, it has played a vital role in our understanding of the abundance of exoplanets in the galaxy. Launched in 2009 by NASA, the Kepler space telescope monitored the light from distant stars, looking for variations that indicated the presence of transiting exoplanets. Their discoveries helped estimate the number of exoplanets in the Milky Way, contributing to our understanding of planetary diversity in our galaxy.

4. Chandra: Exploring the X-ray Universe

The Chandra X-ray Observatory, launched by NASA in 1999, has been vital to the exploration of the Milky Way in the X-ray range. Detecting X-ray emissions from cosmic objects such as neutron stars, black holes and supernova remnants,

Chandra offers a unique view of energetic events and astrophysical processes occurring within our galaxy.

Relevant Earth and Space Observatories

1. Arecibo Observatory: Cutting-Edge Radio Telescope

Located in Puerto Rico, the Arecibo Observatory was a fixed-aperture radio telescope that played a crucial role in observing pulsars, distant galaxies, and the structure of the Milky Way. Unfortunately, the radio telescope collapsed in 2020, but its rich history and contributions to astronomy are remembered as a significant milestone.

2. ALMA: The Atacama Large Millimeter/submillimeter Array

ALMA, located in the Atacama Desert in Chile, is a radio telescope observatory that operates in the millimeter and submillimeter wave bands. With its ability to observe interstellar gas and dust, ALMA is instrumental in studying star formation, protoplanetary disks, and other astrophysical phenomena in the Milky Way.

3. Very Large Telescope (VLT): Cutting-edge Optical Observations

Located in the Atacama Desert in Chile, the Very Large Telescope (VLT) is made up of four interconnected optical telescopes. Operated by the European Southern Observatory (ESO), the VLT is crucial for detailed studies of stars, nebulae and other objects in the Milky Way, providing sharp images and accurate spectra.

4. Mauna Kea Observatories: Rising Above the Clouds

Situated atop the Mauna Kea volcano in Hawaii, these observatories are a collection of ground-based telescopes in a prime location. The high altitude and favorable atmospheric conditions provide a unique opportunity for high-resolution astronomical observations, contributing to significant research into the Milky Way and beyond.

Contributions to the Understanding of the Milky Way.

These ground-based observatories and space missions have been fundamental pillars for expanding our knowledge about the Milky Way. By combining data from different wavelengths and observation techniques, scientists have been able to create a more complete picture of our galaxy, from its overall structure to the details of its individual components.

Future missions, like the James Webb Space Telescope (JWST), are destined to continue this journey of cosmic exploration. By uniting ground-based and space-based observatories, scientists are armed with increasingly advanced tools to unlock the mysteries of the Milky Way, providing valuable insights into the origin, evolution and ongoing dynamics of our galaxy. Each new observation and discovery adds a piece to the cosmic puzzle, bringing us closer to fully understanding the grand spiral we call our galactic home.

CHAPTER 9: LIFE IN THE MILKY WAY: EXPLORING POSSIBILITIES BEYOND EARTH

THE SEARCH FOR EXTRATERRESTRIAL LIFE

The question of the existence of life beyond Earth is one of the most intriguing and fundamental that humanity has set out to explore. In the vastness of the Milky Way, with its billions of stars and countless exoplanets, the search for signs of life has become a cosmic quest, involving astronomical observations, astrobiological studies and interdisciplinary investigations.

1. Exoplanets in the Habitable Zone: Where Life Can Thrive

The search for life in the Milky Way largely focuses on identifying exoplanets in the so-called "habitable zone." This is the region around a star where conditions are right for the existence of liquid water – a vital ingredient for life as we know it. The discovery of exoplanets in this zone raises hope of finding worlds where life could thrive.

2. Theories about the Origin of Life: From Chemistry to Microorganisms

Theories about the origin of life in the Milky Way cover a wide range of possibilities. From the formation of complex organic molecules in interstellar gas clouds to the idea that life may have been transported between planets via comets or meteorites, scientists have explored a range of hypotheses to understand how life could have arisen on exoplanets and, therefore, extension, in our own galaxy.

3. Biomarkers and the Search for Signatures of Life

The search for life in the Milky Way often focuses on detecting "biomarkers" – chemical or physical characteristics that could indicate the presence of life. Among these biomarkers are compounds such as oxygen and methane, which, on Earth, are strongly associated with biological activity. Future space missions, like the James Webb Space Telescope, will look for these signatures in the atmospheres of distant exoplanets.

4. Space Missions Targeted to Search for Extraterrestrial Life

In addition to remote observation of exoplanets, specific space missions are being planned to directly investigate the possibility of extraterrestrial life. The James Webb Space Telescope, with its focus on characterizing exoplanet atmospheres, and future missions such as the Habitable

Exoplanet Space Telescope (HabEx) and the Large UV/Optical/IR Surveyor (LUVOIR), are among the initiatives that could reveal more about the habitability and the existence of life outside our solar system.

5. The Diversity of Exoplanets: Possible Worlds for Life

The diversity of exoplanets discovered so far increases the possibilities of life in the Milky Way. From rocky Earth-like exoplanets to gas giants, each type of world offers a unique environment that could support different forms of life. The discovery of exoplanets in binary and multiple star systems also adds complexity to the potential conditions for extraterrestrial life.

6. The Search for Technological Signals: SETI and Beyond

In addition to the search for biological life forms, scientists are also exploring the possibility of finding technological signs of advanced extraterrestrial civilizations. The SETI (Search for Extraterrestrial Intelligence) project monitors space for radio signals or other transmissions that could indicate the presence of technologically advanced civilizations.

7. Challenges and Limitations of the Search for Extraterrestrial Life

While the search for life in the Milky Way is exciting, it also faces significant challenges. The distance between stars and exoplanets often limits detailed observations, and the complexity of the factors that favor life makes it difficult to predict where and how it might exist. Furthermore, current technological limitations may prevent the detection of certain biomarkers or signs of life.

Conclusion: A Perpetual Search for Life in the Milky Way

The question of the existence of life in the Milky Way is one of the great questions that has echoed since the beginning of humanity. The discovery of exoplanets, the evolution of theories about the origin of life and the advancement of space observation technologies are transforming this search into a solid and focused science.

While we have yet to find concrete evidence of life beyond Earth, continued exploration of the Milky Way, driven by advanced space missions and state-of-the-art ground-based observatories, promises to expand our horizons and bring us ever closer to answering this fundamental cosmic question. . The search for life in the Milky Way is a journey that transcends generations, driven by human curiosity and the tireless quest to

understand our place in the vast and
mysterious cosmos.

CHAPTER 10: CURIOSITIES AND MYTHS: EXPLORING THE CULTURAL NARRATIVES AND INTRIGUING MYSTERIES OF THE MILKY WAY

1. The Milky Way in Greek Mythology: The Milk of Ivy

In Greek mythology, the Milky Way was associated with a legend involving Hera, wife of Zeus, and her son Hercules. According to the story, Hera would have breastfed Hercules while he slept, but when she woke up and realized that he was not her son, she pushed him away, causing her milk to fall across the sky and form the bright band we know as the Milky Way.

2. The Path to Heaven in Different Cultures:

Different cultures have their own interpretations of the Milky Way. In Chinese mythology, the Milky Way is known as the Celestial River, separating the weaver goddess Zhinü (represented by the star Vega) from her lover, the cowherd Niu Lang (represented by the star Altair). Once a year, at a festival known as Qixi, they are allowed to gather.

3. Milky Way in Maori Culture:

The Maori people of New Zealand have their own stories related to the Milky Way. In one narrative, the galaxy is considered to be a canoe, with stars representing the people on board. These stories not only explain what the Milky Way looks like, but also carry deep cultural meanings.

Intriguing Curiosities:

1. Stellar Birth and Death: The Cosmic Dance:

The Milky Way is a stage of constant stellar birth and death. In its spiral arms, nebulae of gas and cosmic dust collapse to form new stars, while other stars age and eventually explode as supernovae. This cosmic dance is one of the driving forces shaping the evolution of our galaxy.

2. Cosmic Spirals and Interstellar Clouds:

The spiral arms of the Milky Way are not only beautiful, but they are also active sites of star formation. Interstellar clouds, composed of gas and dust, are the cosmic nurseries where new stars are born. These spiral arms serve as fascinating clues to the dynamics and evolution of the Milky Way.

3. Galactic Halo and Ancient Stars:

Beyond the galactic disk, the Milky Way is surrounded by a halo of ancient stars. These stars form a distinct stellar population and provide important clues about the events that shaped the early days of our galaxy. Studying the galactic halo is like leafing through the pages of a cosmic book that tells the story of the Milky Way.

4. The Black Hole in the Center: Sagittarius A*:

At the core of the Milky Way, hidden in the constellation Sagittarius, resides a supermassive black hole known as Sagittarius A*. This black hole exerts a significant gravitational influence, affecting the dynamics of the stars around it. Studying Sagittarius A* offers crucial insights into the nature of black holes and their influence on the galaxy.

5. The Colliding Milky Way: A Future Cosmic Dance:

The Milky Way is on a collision course with the Andromeda galaxy, our cosmic neighbor. Although this merger is billions of years away, it represents an intriguing glimpse into the cosmic future. The collision will create a new elliptical galaxy, mixing stars, gas and dust in a spectacular cosmic dance.

6. The Mystery of Dark Matter and Dark Energy:

Most of the Milky Way's mass is made up of dark matter, a mysterious substance that does not interact with light and has not yet been directly detected. Furthermore, dark energy, a force that drives the accelerating expansion of the universe, remains one of the greatest mysteries in astrophysics. Both of these components add layers of complexity to understanding the Milky Way.

7. Cosmic Speed:

Earth travels at incredible speed around the center of the Milky Way, completing one orbit every 225 million years. This means that every human being experiences an extraordinary cosmic journey throughout their lives, traveling vast distances in galactic space.

8. The Sound of the Milky Way:

The Milky Way emits radio waves that, when converted into audio, produce a distinct sound. This sound, known as "galactic background noise", is a cosmic symphony of emissions from stars, gas and other objects in our galaxy.

Conclusion: A Cosmic Kaleidoscope of Myths and Wonders

The Milky Way, our galaxy, is not just an immense cosmic structure, but a kaleidoscope of cultural myths and cosmic wonders. Over the centuries, different civilizations have attributed different meanings to the glowing band that adorns our night sky. Simultaneously, modern science continues to unravel the intrinsic mysteries of the Milky Way, revealing its unparalleled complexity and beauty.

As we explore the mythologies that have enriched our cultural understanding of the Milky Way and delve into the fascinating curiosities that science reveals, we are reminded of the vastness of the universe and the human capacity to question, imagine, and unlock the deepest secrets of the cosmos. The Milky Way remains not just an astronomical entity, but an inexhaustible source of inspiration, reflection, and discovery in our ongoing quest for cosmic knowledge.

CHAPTER 11: PHOTOGRAPHS AND SPACE IMAGES: A VISUAL JOURNEY THROUGH THE MILKY WAY

EXPLORING THE MILKY WAY THROUGH THE COSMIC LENS

The Milky Way, with its vastness and splendor, is an inexhaustible source of inspiration for astronomers, astrophotographers and cosmos enthusiasts. In this chapter, we will embark on a visual journey, exploring stunning images that capture the unique beauty of our galaxy, as well as the fascinating celestial objects that populate it.

1. The Milky Way Stellar Mosaic:

Let's start our visual journey with a stellar mosaic of the Milky Way, captured by space telescopes like Hubble and ground-based observatories. This panoramic image reveals the extent of the galactic disk, with its spiral arms adorned by a myriad of stars.

2. Star Formation Nebulae: Cosmic Nurseries:

Among the most captivating images of the Milky Way are those that show us star-forming nebulae. The Orion Nebula, within our galaxy, is a notable example. Its clouds of cosmic gas and dust are illuminated by young stars, creating a spectacle of stunning colors and patterns.

3. The Spiral Arms of the Milky Way: A Cosmic Dance:

An image that reveals the spiral structure of the Milky Way is like a work of cosmic art. Panoramic photographs of the spiral arms, highlighting regions of intense star formation, provide a unique view of galactic dynamics.

4. The Galactic Nucleus and Sagittarius A*:

The central region of the Milky Way, where the supermassive black hole Sagittarius A* resides, is an intriguing target for observations. Long-exposure photographs reveal stars orbiting the black hole, creating dizzying patterns around the focal point.

5. Star Clusters: Concentrations of Cosmic Light:

The Milky Way is home to numerous star clusters, where stars group together due to their common formation. Photographs of these clusters, such as the Wild Duck Star Cluster, highlight the diversity of stars present in the galaxy.

6. Binary Black Holes: Dancing in the Dark:

The detection of binary black holes in the Milky Way is a remarkable feat. Images capturing these systems, in which two black holes orbit each other, offer glimpses of the extreme gravitational interaction taking place deep in space.

7. The Zodiacal Light and the Earth's Atmosphere:

Photographs of the Milky Way often incorporate terrestrial elements such as zodiacal light, which is sunlight reflected by dust particles in the plane of the solar system. This combination of celestial and earthly elements creates visually stunning compositions.

8. The Aurora Phenomenon and the Milky Way:

In regions close to the poles, the interaction of solar particles with the Earth's atmosphere creates stunning auroras. Photographs that capture auroras alongside the Milky Way present a unique view of the interaction between the cosmos and Earth.

ASTROPHOTOGRAPHER
GÖRAN STRAND
astrofotografen.se

9. Transiting Exoplanets: Small Points of Interest:

Although it is not possible to photograph exoplanets directly, observations such as stellar transits can reveal small variations in a star's light caused by the passing of a planet. These indirect observations are a window into the diversity of worlds orbiting stars in the Milky Way.

10. The Galactic Merger with Andromeda: The Cosmic Future:

Imagining the cosmic future of the Milky Way involves visualizing the future collision with the Andromeda galaxy. Artistic representations based on scientific simulations offer us an intriguing vision of this cosmic spectacle that will take place in billions of years.

Conclusion: A Visual Odyssey through the Milky Way

These images, captured by ground-based observatories and space telescopes, provide a visual odyssey through the Milky Way, highlighting its stunning beauty and the diversity of cosmic phenomena within it. Each photograph is more than a simple image; it is a portal that connects us with the wonders of the galaxy to which we belong.

As we continue to advance our technological capabilities and explore the far reaches of the Milky Way, we can expect even more spectacular images that will not only leave us in awe of the vastness of the universe, but also inspire us to seek answers to the cosmic mysteries that still await discovery. The Milky Way, in all its visual glory, stands as a silent and majestic witness to the grandeur of the cosmos.

CHAPTER 12: CHALLENGES AND UNSOLVED MYSTERIES:

The search for cosmic knowledge is a constant journey, filled with intriguing challenges and mysteries that continue to defy scientific understanding. In this chapter, we will explore some of the most challenging areas and mysterious phenomena that persist in the Milky Way, revealing the unexplored frontiers of astronomy and physics.

1. Dark Matter and Dark Energy: The Invisible Shadows

One of the biggest enigmas in contemporary astrophysics lies in the mysterious forms of dark matter and dark energy. Dark matter, although it exerts gravitational influence, does not emit, absorb or reflect light, making it invisible to traditional observation methods. Dark energy, on the other hand, is responsible for the accelerated expansion of the universe. Both make up most of the contents of the cosmos, but their true nature remains elusive, challenging scientists to develop innovative methods to detect and understand them.

2. Origin of Life: From Cosmic to Microscopic

Although we have a reasonable understanding of the physical and chemical processes that sustain life on Earth, the origin of life remains an intriguing mystery. How did the complex organic molecules that eventually gave rise to living beings emerge? The search for answers involves exploring conditions on planets, moons and even comets, examining the conditions necessary for the creation of life and its possible spread throughout the cosmos.

3. Existence of Habitable Exoplanets: Searching for Earth-Like Worlds

Finding exoplanets in the so-called "habitable zone" is a crucial objective in the search for extraterrestrial life. However, identifying planets that are not only in the habitable zone, but also have favorable conditions for life, is a complex challenge. Understanding the atmospheres of these exoplanets and detecting biomarkers are active areas of research, with telescopes like the James Webb Space Telescope (JWST) poised to provide valuable information.

4. The Nature of Black Holes: Riddles in the Heart of Darkness

Black holes continue to challenge conventional understanding of physics. The singularity at the center of a black hole, where the known laws of physics appear to fail, is a fundamental question mark. Furthermore, the connection between the theory of general relativity and quantum mechanics in extreme environments, such as near an event horizon, is an open mystery.

5. The Physics of Gravity: Extreme Tests in Gravitational Fields

Although Einstein's theory of general relativity is a remarkably accurate description of gravity, it is mainly tested in moderate gravitational fields. In more extreme environments, such as near black holes or during supernova explosions, the laws of gravity can behave in unexpected ways. Testing the physics of gravity in these extreme conditions is an important challenge that could lead to a deeper understanding of the fundamental nature of the universe.

6. Nature of High-Energy Cosmic Rays: Mystical Messengers from Space

High-energy cosmic rays are extremely energetic subatomic particles that travel through space at speeds approaching that of light. The origin of these particles, their trajectories through the cosmos, and the impact they have on our environment are still topics of active investigation. Ground-based detectors and space observatories are used to study these mystical cosmic messengers, but many details remain enigmatic.

7. Galactic Fusion and the Fate of the Milky Way: A Cosmic Prediction

While we have a general prediction that the Milky Way is destined to collide with the Andromeda galaxy, the precise details of this cosmic merger are still the subject of study. Computational models and simulations are crucial tools for understanding the effects of this collision on our stellar systems, spiral arms and even the possibility of new star formation.

8. The Role of Neutron Stars and Quasars in the Galaxy: Beyond Known Limits

Neutron stars and quasars are extreme cosmic phenomena, and their role in galactic dynamics is still not entirely clear. How do these massive objects affect the structure and evolution of the Milky Way? Studying these cosmic events requires combining high-resolution observations and sophisticated theoretical models.

9. Exoplanet Observation: More Than What We See Now

As we advance the observation of exoplanets, there are ongoing challenges, such as the need to develop more refined techniques for characterizing the atmospheres of these distant worlds. The direct detection of signs of life, as biomarkers, also remains an ambitious and complex objective, requiring technological innovations and advances in observation methods.

10. The Nature of Cosmic Darkness: Unlocking the Secrets of the Invisible Universe

The cosmic darkness, made up of dark matter and dark energy, continues to challenge scientists. With the search for direct detections of dark matter and deeper studies into the nature of dark energy, astronomers are determined to unlock the secrets of these invisible components that dominate the composition of the universe.

Conclusion: Towards the Cosmic Unknown

As we explore the Milky Way and beyond, the scientific journey is marked by exciting challenges and intriguing mysteries. Each unsolved mystery represents a unique opportunity to advance our understanding of the universe and discover the truths hidden in the vast cosmic reaches. Science continues its unremitting quest to unravel these enigmas, pushing the frontiers of human knowledge beyond what is currently understood. The cosmic unknown, full of promise and revelation, remains an irresistible invitation for curious explorers of the cosmos.

CHAPTER 13: SCIENTIFIC AND CULTURAL IMPORTANCE OF THE MILKY WAY: A JOURNEY OF DISCOVERY AND INSPIRATION

SCIENTIFIC CONTRIBUTIONS:

The Milky Way, our galactic home, plays a crucial role in understanding the universe and expanding scientific knowledge. The importance of studying our galaxy goes beyond the mere desire to know the place we call our cosmic home; it opens doors to a deeper understanding of the fundamental principles that govern the cosmos.

1. Contextualization of the Universe:

By studying the Milky Way, we not only learn about our galaxy, but we also gain valuable insights into the structure and evolution of galaxies in general. The diversity of stars, planetary systems, black holes and other objects found in the Milky Way provides a rich and varied picture that contributes to the development of theories about the formation and evolution of stellar and galactic systems throughout the universe.

2. Cosmology and Origins:

The Milky Way is a testimony to the vastness of space and time. By studying the distribution of stars in different regions of the galaxy, scientists can infer the relative ages of these stars and, by extension, develop an approximate timeline for the formation of the Milky Way. This information is essential for understanding the origins of the universe and the processes that gave rise to our galaxy and others.

3. Search for Extraterrestrial Life:

The investigation of exoplanets in the Milky Way plays a fundamental role in the search for extraterrestrial life. By identifying planets in the habitable zone and characterizing their atmospheres, astronomers can assess the possibility of conditions conducive to life. This research, although still in its early stages, may eventually provide clues about the prevalence of life in the universe.

4. Extreme Physics:

Studying extreme phenomena in the Milky Way, such as black holes and neutron stars, offers a unique opportunity to test fundamental theories of physics under extreme conditions. The intense gravity near a black hole, for example, provides a cosmic laboratory for investigating the interactions between general relativity and quantum mechanics.

Cultural and Artistic Influence:

In addition to scientific contributions, the Milky Way has also exerted a significant influence on human culture throughout history. Whether as a source of artistic inspiration or as a central element in mythologies and cultural narratives, the galaxy has played a vital role in the human imagination.

1. Mythology and Narratives:

Numerous cultures around the world have incorporated the Milky Way into their mythologies. In Ancient Greece, the galaxy was associated with Hera's spilled milk, while in China it was seen as the Celestial River separating divine lovers. These narratives not only provide an explanation for the band of light in the night sky, but also reflect the human capacity to attribute cosmic meaning to natural phenomena.

2. Artistic Inspiration:

Throughout history, artists have been inspired by the beauty of the Milky Way to create captivating works of art. Paintings, photographs and artistic renderings capture the majesty of the spiral arms, the luminosity of the galactic core and the cosmic poetry that the galaxy evokes. The Milky Way becomes a muse that transcends cultural boundaries, uniting people through a shared appreciation for the wonder of the cosmos.

3. Celestial Navigation:

Before the advent of satellite navigation, the stars of the Milky Way were used as guides for navigation. Many ancient cultures and indigenous peoples used the Milky Way's position in the sky for nighttime orientation, marking seasons and determining directions. The galaxy, therefore, played a practical and spiritual role in the relationship between societies and space.

4. Space Exploration:

The exploration of the Milky Way also sparked a fascination with space exploration. As space missions provide us with increasingly detailed images and information about our galaxy, public interest in deep space exploration continues to grow. The Milky Way becomes not only an object of scientific study, but also a source of inspiration for new generations of cosmic explorers.

CHAPTER 14: CONCLUSION - NAVIGATING THE MYSTERIES OF THE MILKY WAY

Along this journey through the Milky Way, we delve into the depths of the cosmos, exploring its wonders, challenges and mysteries. The galaxy we call home has proven to be more than just a collection of stars; It is a dynamic stage where science and culture intertwine, where the unknown awaits discovery, and where cosmic beauty inspires the human imagination.

A Universe in Our Arms:

The Milky Way, with its spiral arms, planetary systems and profound mysteries, reveals the unparalleled complexity and diversity of the cosmos. Through studying their structures and compositions, we have not only unlocked the secrets of our galaxy, but also gained fundamental insights into the nature of galaxies in general.

Scientific Challenges and Intriguing Mysteries:

We explore scientific challenges, from the search for dark matter to understanding the origin of life and the nature of black holes. Faced with these cosmic enigmas, we realize that the Milky Way is a celestial laboratory where the laws of physics are tested in the extreme conditions of the universe.

Culture, Art and Inspiration:

Beyond the scientific realm, the Milky Way has woven its influence into the tapestry of human culture. From ancient mythologies to contemporary artistic inspirations, the galaxy has transcended its astronomical role, becoming a source of meaning, reflection and inspiration.

The Continuity of Exploration:

Exploration of the Milky Way continues. New space missions, technological advances and surprising discoveries promise to broaden our horizons and reveal deeper layers of knowledge about our galaxy and the universe it inhabits.

The Milky Way: An Invitation to Discovery:

Ultimately, the Milky Way remains an invitation to continued discovery. Its history, structure, and mysteries await those who seek to understand not only the secrets of the cosmos, but also the galaxy's role in our understanding of the universe.

Looking to the Stellar Future:

As we continue our exploration of space, may this book serve as a starting point for your own cosmic journey. The Milky Way, with its immensity and beauty, is a silent witness to the vastness of the universe, inviting us to dream, explore and question.

May each star in the Milky Way be a reminder that the unknown lies ahead of us, and may each mystery be an opportunity to unlock the complexities of the cosmos. As we say goodbye to this journey, we look up at the night sky with a new understanding and a deep appreciation for the galaxy we call home.

The adventure through the Milky Way continues, and the stars whisper ancient stories, inspiring us to lift our eyes and seek the cosmic infinity that stretches beyond what we can imagine. May each future discovery bring us one step closer to fully understanding this luminous spiral that envelops us in the vast universe.

www.ingramcontent.com/pod-product-compliance
Lightning Source LLC
Chambersburg PA
CBHW071056290526
45795CB00004B/1527